1 MONTH OF
FREE
READING

at

www.ForgottenBooks.com

By purchasing this book you are eligible for one month membership to ForgottenBooks.com, giving you unlimited access to our entire collection of over 1,000,000 titles via our web site and mobile apps.

To claim your free month visit:

www.forgottenbooks.com/free113590

ISBN 978-1-5283-7693-8

PIBN 10113590

Historic, Archive Document

Do not assume content reflects current scientific policies, or practices.

U. S. DEPARTMENT OF AGRICULTURE.

FARMERS' BULLETIN No. 54.

SOME COMMON BIRDS

IN THEIR RELATION TO AGRICULTURE.

BY

F. E. L. BEAL, B. S.,

ASSISTANT ORNITHOLOGIST, BIOLOGICAL SURVEY.

[May, 1897.]

WASHINGTON:

GOVERNMENT PRINTING OFFICE,

1897.

CONTENTS

2

SOME COMMON BIRDS IN THEIR RELATION TO AGRICULTURE.

INTRODUCTION.

It has long been known that birds play an important part in relation to agriculture, but there seems to be a tendency to dwell on the harm they do rather than on the good. Whether a bird is injurious or beneficial depends almost entirely upon what it eats, and in the case of species which are unusually abundant or which depend in part upon the farmer's crops for subsistence the character of the food often becomes a very practical question. If crows or blackbirds are seen in numbers about cornfields, or if woodpeckers are noticed at work in an orchard, it is perhaps not surprising that they are accused of doing harm. Careful investigation, however, often shows that they are actually destroying noxious insects, and also that even those which do harm at one season may compensate for it by eating noxious species at another. Insects are eaten at all times by the majority of land birds, and during the breeding season most kinds subsist largely and rear their young exclusively on this food. When insects are unusually plentiful, they are eaten by many birds which ordinarily do not touch them. Even birds of prey resort to this diet, and when insects are more easily obtained than other fare, the smaller hawks and owls live on them almost entirely. This was well illustrated during the recent plague of Rocky Mountain locusts in the Western States, when it was found that locusts were eaten by nearly every bird in the region, and that they formed almost the entire food of a large majority of the species.

Within certain limits, birds feed upon the kind of food that is most accessible. Thus, as a rule, insectivorous birds eat the insects that are most easily obtained, provided they do not have some peculiarly disagreeable property. It is not probable that a bird habitually passes by one kind of insect to look for another which is more appetizing, and there seems little evidence in support of the theory that the selection of food is restricted to any particular species of insect, for it is evident that a bird eats those which by its own method of seeking are most easily obtained. Thus, a ground-feeding bird eats those it finds among the dead leaves and grass; a flycatcher, watching for its prey from some vantage point, captures entirely different kinds; and the woodpecker and warbler, in the tree tops, select still others. It is thus apparent that a bird's diet is likely to be quite varied, and to differ at different seasons of the year.

In investigating the food habits of birds, field observation can be relied on only to a limited extent, for it is not always easy to determine what a bird really eats by watching it. In order to be positive on this point, it is necessary to examine the stomach contents. When birds are suspected of doing injury to field crops or fruit trees, a few individuals should be shot and their stomachs examined. This will show unmistakably whether or not the birds are guilty.

In response to a general demand for definite information regarding the food habits of our native birds, the Biological Survey of the Department of Agriculture has for some years past been conducting a systematic investigation of the food of species which are believed to be of economic importance. Thousands of birds' stomachs have been carefully examined in the laboratory, and all the available data respecting the food brought together. The results of the investigations relating to birds of prey, based on an examination of nearly 3,000 stomachs, were published in 1893, in a special bulletin entitled The Hawks and Owls of the United States. Many other species have been similarly studied and the results published, either in special bulletins or as articles in the yearbooks. The present bulletin contains brief abstracts of the results of food studies of about 30 grain and insect eating birds belonging to 10 different families.[1]

These species comprise among others the crow blackbirds and ricebirds, against which serious complaints have been made on account of the damage they do to corn, wheat, rice, and other crops; and also the cuckoos, grosbeaks, and thrashers, which are generally admitted to be beneficial, but whose true value as insect destroyers has not been fully appreciated. The practical value of birds in controlling insect pests should be more generally recognized. It may be an easy matter to exterminate the birds in an orchard or grain field, but it is an extremely difficult one to control the insect pests. It is certain, too, that the value of our native sparrows as weed destroyers is not appreciated. Weed seed forms an important item of the winter food of many of these birds, and it is impossible to estimate the immense numbers of noxious weeds which are thus annually destroyed.

[1] The limits of this bulletin preclude giving more than a very brief statement regarding the food of each bird, but more detailed accounts of some of the species will be found in the following reports of the Biological Survey (formerly Division of Ornithology and Mammalogy): The Crow—Bulletin No. 6, 1895, pp. 1-98; Woodpeckers—Bulletin No. 7, 1895, pp. 1-39; Kingbird—Annual Report Secretary of Agriculture, for 1893, pp. 233-234; Baltimore Oriole—Yearbook United States Department of Agriculture for 1895, pp. 426-430; Grackles—Yearbook for 1894, pp. 233-248; Meadowlark—Yearbook for 1895, pp. 420-426; Cedarbird—Annual Report Secretary of Agriculture, for 1892, pp. 197-200; Catbird, Brown Thrasher, and Wren—Yearbook for 1895, pp. 405-418.

If birds are protected and encouraged to nest about the farm and garden, they will do their share in destroying noxious insects and weeds, and a few hours spent in putting up boxes for bluebirds, martins, and wrens will prove a good investment. Birds are protected by law in many States, but it remains for the agriculturists to see that the laws are faithfully observed.

THE CUCKOOS.

(Coccyzus americanus and C. erythrophthalmus.)

Two species of cuckoos, the yellow-billed (fig. 1) and the black-billed, are common in the United States east of the Plains, and a subspecies of the yellow-billed extends westward to the Pacific. While the two species are quite distinct, they do not differ greatly in food habits, and their economic status is practically the same.

Fig. 1.—Yellow-billed cuckoo.

An examination of 37 stomachs has shown that these cuckoos are much given to eating caterpillars, and, unlike most birds, do not reject those covered with hair. In fact, cuckoos eat so many hairy caterpillars that the hairs pierce the inner lining of the stomach and remain there, so that when the stomach is opened and turned inside out, it appears to be lined with a thin coating of fur.

An examination of the stomachs of 16 black-billed cuckoos, taken during the summer months, showed the remains of 328 caterpillars, 11 beetles, 15 grasshoppers, 63 sawflies, 3 stink bugs, and 4 spiders. In all probability more individuals than these were represented, but their remains were too badly broken for recognition. Most of the caterpillars were hairy, and many of them belonged to a genus that lives in colonies and feeds on the leaves of trees, including the apple tree. One stomach was filled with larvæ of a caterpillar belonging to the same genus as the tent caterpillar, and possibly to that species. Other larvæ were those of large moths, for which the bird seems to have a special fondness. The beetles were for the most part click beetles and weevils, with a few May beetles and some others. The sawflies were all found in two stomachs, one of which contained no less than 60 in the larval stage.

Of the yellow-billed cuckoo, 21 stomachs (collected from May to October, inclusive) were examined. The contents consisted of 355 caterpillars, 18 beetles, 23 grasshoppers, 31 sawflies, 14 bugs, 6 flies, and 12 spiders. As in the case of the black-billed cuckoo, most of the caterpillars belonged to hairy species and many of them were of large size. One stomach contained 12 American tent caterpillars; another 217 fall webworms. The beetles were distributed among several families, but all more or less harmful to agriculture. In the same stomach which contained the tent caterpillars were two Colorado potato beetles; in another were three goldsmith beetles and remains of several other large beetles. Besides ordinary grasshoppers were several katy-dids and tree crickets. The sawflies were in the larval stage, in which they resemble caterpillars so closely that they are commonly called false caterpillars by entomologists, and perhaps this likeness may be the reason the cuckoos eat them so freely. The bugs consisted of stink bugs and cicadas or dog-day harvest flies, with the single exception of one wheel bug, which was the only useful insect eaten, unless the spiders be counted as such.

THE WOODPECKERS.

Five or six species of woodpeckers are familiarly known throughout the eastern United States, and in the west are replaced by others of similar habits. Several species remain in the northern States through the entire year, while others are more or less migratory.

Farmers are prone to look upon woodpeckers with suspicion. When the birds are seen scrambling over fruit trees and pecking at the bark, and fresh holes are found in the tree, it is concluded that they are doing harm. Careful observers, however, have noticed that, excepting a single species, these birds rarely leave any important mark on a healthy tree, but that when a tree is affected by wood-boring larvæ the insects are accurately located, dislodged, and devoured. In case the holes from which the borers are taken are afterwards occupied and enlarged by colonies of ants, these ants in turn are drawn out and eaten.

Two of the best known woodpeckers, the hairy woodpecker (*Dryobates villosus*) (fig. 2) and the downy woodpecker (*D. pubescens*), including their races, range over the greater part of the United States, and for the most part remain throughout the year in their usual haunts. They differ chiefly in size, for their colors are practically the same, and the males, like other woodpeckers, are distinguished by a scarlet patch on the head.

An examination of many stomachs of these two birds shows that from two-thirds to three-fourths of the food consists of insects, chiefly noxious. Wood-boring beetles, both adults and larvæ, are conspicuous, and with them are associated many caterpillars, mostly species that burrow into trees. Next in importance are the ants that live in decaying wood, all of which are sought by woodpeckers and eaten in great quantities. Many ants are particularly harmful to timber, for if they find a small spot of decay in the vacant burrow of some woodborer, they enlarge the hole, and as their colony is always on the increase, continue to eat away the wood until the whole trunk is honey-

Fig. 2.—Hairy woodpecker.

combed. Moreover, these insects are not accessible to other birds, and could pursue their career of destruction unmolested were it not that the woodpeckers, with beaks and tongues especially fitted for such work, dig out and devour them. It is thus evident that woodpeckers are great conservators of forests. To them, more than to any other agency, we owe the preservation of timber from hordes of destructive insects.

One of the larger woodpeckers familiar to everyone is the flicker, or golden-winged woodpecker (*Colaptes auratus*) (fig. 3), which is generally distributed throughout the United States from the Atlantic Coast to the Rocky Mountains. It is there replaced by the red-shafted flicker (*C. cafer*), which extends westward to the Pacific. The two species are as nearly identical in food habits as their environment will allow. The flickers, while genuine woodpeckers, differ somewhat in habits from the rest of the family, and are frequently seen upon the ground searching for food. Like the downy and hairy woodpeckers, they eat wood-boring grubs and ants, but the number of ants eaten is much greater. Two of the flickers' stomachs examined were completely filled with ants, each stomach containing more than 3,000 individuals. These ants belonged to species which live in the ground, and it is these insects for which the flicker is searching when running about in the grass, although some grasshoppers are also taken.

Fig. 3.—Flicker.

The red-headed woodpecker (*Melanerpes erythrocephalus*) (fig. 4) is well known east of the Rocky Mountains, but is rather rare in New England. Unlike some of the other species, it prefers fence posts and telegraph poles to trees as a foraging ground. Its food therefore naturally differs from that of the preceding species, and consists largely of adult beetles and wasps, which it frequently captures on the wing, after the fashion of flycatchers. Grasshoppers also form an important part of the food. The redhead has a peculiar habit of selecting very large beetles, as shown by the presence of fragments of several of the largest species in the stomachs. Among the beetles were quite a number of predaceous ground beetles, and unfortunately some tiger beetles, which are useful insects. The redhead has been accused of robbing the nests of other birds; also of attacking young birds and poultry and pecking out their brains, but as the stomachs showed little evidence to substantiate this charge it is probable that the habit is rather exceptional.

It has been customary to speak of the smaller woodpeckers as "sapsuckers," under the belief that they drill holes in the bark of trees for the purpose of drinking the sap and eating the inner bark. Close observation, however, has fixed this habit upon only one species, the yellow-bellied woodpecker, or sapsucker (*Sphyrapicus varius*) (fig 5). This bird has been shown to be guilty of pecking holes in the bark of various forest trees, and sometimes in that of apple trees, from which it drinks the sap

FIG 4.—Red-headed woodpecker.

when the pits become filled. It has been proved, however, that besides taking the sap the bird captures large numbers of insects which are attracted by the sweet fluid, and that these form a very considerable portion of its diet. In some cases the trees are injured by being thus punctured, and die in a year or two, but since comparatively few are touched the damage is not great. It is equally probable, moreover, that the bird fully compensates for this injury by the insects it consumes.

The vegetable food of woodpeckers is varied, but consists largely of small fruits and berries. The downy and hairy woodpeckers eat such fruits as dogwood, Virginia creeper, and others, with the seeds of poison ivy, sumac, and a few other shrubs. The flicker also eats a great many small fruits and the seeds of a considerable number of shrubs and weeds. None of the three species is much given to eating cultivated fruits or crops.

The redhead has been accused of eating the larger kinds of fruit, such as apples, and also of taking considerable corn. The stomach examinations show that to some extent these charges are substantiated, but that the habit is not prevalent enough to cause much damage. It is quite fond of mast, especially beechnuts, and when these nuts are

Fig. 5.—Yellow-bellied woodpecker.

plentiful the birds remain north all winter, instead of migrating as is their usual custom.

Half the food of the sapsucker, aside from sap, consists of vegetable matter, largely berries of the kinds already mentioned, and also a quantity of the inner bark of trees, more of which is eaten by this species than by any other.

Many other woodpeckers are found in America, but their food habits agree in the main with those just described. These birds are certainly the only agents which can successfully cope with certain insect enemies of the forests, and, to some extent, of fruit tree salso. For this reason, if for no other, they should be protected in every possible way.

THE KINGBIRD.

(*Tyrannus tyrannus.*)

The kingbird (fig. 6) is essentially a lover of the orchard, and wherever the native groves have been replaced by fruit trees this pugnacious bird takes up its abode. It breeds in all of the States east of the Rocky Mountains, and less commonly in the Great Basin and on the Pacific Coast. It migrates south early in the fall, and generally leaves the United States to spend the winter in more southern latitudes.

FIG. 6.—Kingbird.

The kingbird manifests its presence in many ways. It is somewhat boisterous and obtrusive, and its antipathy for hawks and crows is well known. It never hesitates to give battle to any of these marauders, no matter how superior in size, and for this reason a family of kingbirds is a desirable adjunct to a poultry yard. On one occasion in the knowledge of the writer a hawk which attacked a brood of young turkeys was pounced upon and so severely buffeted by a pair of kingbirds, whose nest was near by, that the would-be robber was glad to escape without his prey. Song birds that nest near the kingbird are similarly protected.

In its food habits this species is largely insectivorous. It is a true flycatcher by nature, and takes a large part of its food on the wing. It does not, however, confine itself to this method of hunting, but picks up some insects from trees and weeds, and even descends to the ground in search of myriapods or thousand legs. The chief complaint against the kingbird is that it preys largely upon honeybees; and this charge has been made both by professional bee keepers and others. Many observers have seen the bird at work near hives, and there is no reason to doubt the honesty of their testimony. One bee raiser in Iowa, suspecting the kingbirds of feeding upon his bees, shot a number near his hives, but when the birds' stomachs were examined by an expert entomologist not a trace of honeybees could be found.

The Biological Survey has made an examination of 281 stomachs collected in various parts of the country, but found only 14 containing remains of honeybees. In these 14 stomachs there were in all 50 honeybees, of which 40 were drones, 4 were certainly workers, and the remaining 6 were too badly broken to be identified as to sex.

The insects that constitute the great bulk of the food of this bird are noxious species, largely beetles—May beetles, click beetles (the larvæ of which are known as wire worms), weevils, which prey upon fruit and grain, and a host of others. Wasps, wild bees, and ants are conspicuous elements of the food, far outnumbering the hive bees. During summer many grasshoppers and crickets, as well as leaf hoppers and other bugs, are also eaten. Among the flies were a number of robber flies—insects which prey largely upon other insects, especially honeybees, and which have been known to commit in this way extensive depredations. It is thus evident that the kingbird by destroying these flies actually does good work for the apiarist. Nineteen robber flies were found in the stomachs examined; these may be considered more than an equivalent for the four worker honeybees already mentioned. A few caterpillars are eaten, mostly belonging to the group commonly known as cutworms, all the species of which are harmful. About 10 per cent of the food consists of small native fruits, comprising some twenty common species of the roadsides and thickets, such as dogwood berries, elder berries, and wild grapes. The bird has not been reported as eating cultivated fruit to an injurious extent, and it is very doubtful if this is ever the case, for cherries and blackberries are the only ones that might have come from cultivated places, and they were found in but few stomachs.

Three points seem to be clearly established in regard to the food of the kingbird—(1) that about 90 per cent consists of insects, mostly injurious species; (2) that the alleged habit of preying upon honeybees is much less prevalent than has been supposed, and probably does not result in any great damage; and (3) that the vegetable food consists almost entirely of wild fruits which have no economic value. These facts, taken in connection with its well-known enmity for hawks and crows, entitle the kingbird to a place among the most desirable birds of the orchard or garden.

THE PHŒBE.

(Sayornis phœbe.)

Among the early spring arrivals at the North, none are more welcome than the phœbe (fig. 7). Though naturally building its nest under an overhanging cliff of rock or earth, or in the mouth of a cave, its preference for the vicinity of farm buildings is so marked that in the more thickly settled parts of the country the bird is seldom seen at any great distance from a farmhouse except where a bridge spans some stream, affording a secure spot for a nest. Its confiding disposition has rendered it a great favorite, and consequently it is seldom disturbed. It breeds throughout the United States east of the Great Plains, and winters from the South Atlantic and Gulf States southward.

Fig. 7.—Phœbe.

The phœbe subsists almost exclusively upon insects, most of which are caught upon the wing. An examination of 80 stomachs showed that over 93 per cent of the year's food consists of insects and spiders, while wild fruit constitutes the remainder. The insects belong chiefly to noxious species, and include many click beetles, May beetles, and weevils. Grasshoppers in their season are eaten to a considerable extent, while wasps of various species, many flies of species that annoy cattle, and a few bugs and spiders are also eaten regularly. It is evident that a pair of phœbes must materially reduce the number of insects near a garden or field, as the birds often, if not always, raise two broods a year, and each brood numbers from four to six young.

The vegetable portion of the food is unimportant, and consists mainly of a few seeds, with small fruits, such as wild cherries, elder berries, and juniper berries. The raspberries and blackberries found in the stomachs were the only fruits that might have belonged to cultivated varieties, and the quantity was trifling.

There is hardly a more useful species than the phœbe about the farm, and it should receive every encouragement. To furnish nesting boxes is unnecessary, as it usually prefers a more open situation, like a shed, or a nook under the eaves, but it should be protected from cats and other marauders.

THE BLUE JAY.

(*Cyanocitta cristata.*)

The blue jay (fig. 8) is a common bird of the United States east of the Great Plains, and remains throughout the year in most of its range, although its numbers are somewhat reduced in winter in the Northern States. During spring and summer the jay is forced to become an

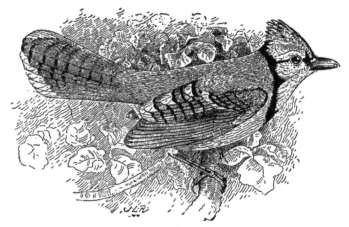

Fig. 8.—Blue jay.

industrious hunter for insects, and is not so conspicuous a feature of the landscape as when it roams the country at will after the cares of the nesting season are over.

Ornithologists and field observers in general declare that a considerable portion of its food in spring and early summer consists of the eggs and young of small birds, and some farmers accuse it of stealing corn to an injurious extent in the fall. While there may be some truth in these accusations, they have almost certainly been exaggerated. No doubt many jays have been observed robbing nests of other birds, but thousands have been seen that were not so engaged.

In an investigation of the food of the blue jay 292 stomachs were examined, which showed that animal matter comprised 24 per cent and vegetable matter 76 per cent of the bird's diet. So much has been said about the nest-robbing habits of the jay that special search was made for traces of birds or birds' eggs in the stomachs, with the result that shells of small birds' eggs were found in three and the remains of young birds in only two stomachs. Such negative evidence is not sufficient to controvert the great mass of testimony upon this point, but it shows that the habit is not so prevalent as has been believed. Besides birds and their eggs, the jay eats mice, fish, salamanders, snails, and crustaceans, which altogether constitute but little more than 1 per cent of its diet. The insect food is made up of beetles, grasshoppers, caterpillars, and a few species of other orders, all noxious, except some $3\frac{1}{2}$ per cent of predaceous beetles. Thus something more than 19 per cent of the whole food consists of harmful insects. In August the jay, like many other birds, turns its attention to grasshoppers, which constitute nearly one-fifth of its food during that month. At this time, also, most of the other noxious insects, including caterpillars, are consumed, though beetles are eaten chiefly in spring.

The vegetable food is quite varied, but the item of most interest is grain. Corn was found in 70 stomachs, wheat in 8, and oats in 2—all constituting 19 per cent of the total food. Corn is evidently the favorite grain, but a closer inspection of the record shows that the greater part was eaten during the first five months of the year, and that very little was taken after May, even in harvest time, when it is abundant. This indicates that most of the corn is gleaned from the fields after harvest, except what is stolen from cribs or gathered in May at planting time.

The jay's favorite food is mast (i. e., acorns, chestnuts, chinquapins, etc.), which was found in 158 of the 292 stomachs and amounted to more than 42 per cent of the whole food. In September corn formed 15 and mast 35 per cent, while in October, November, and December corn dropped to an almost inappreciable quantity and mast amounted to 64, 82, and 83 per cent, respectively. And yet in these months corn is abundant and everywhere easily accessible. The other elements of food consist of a few seeds and wild fruits, among which grapes and blackberries predominate.

The results of the stomach examination show, (1) that the jay eats many noxious insects; (2) that its habit of robbing the nests of other birds is much less common than has been asserted; and (3) that it does little harm to agriculture, since all but a small amount of the corn eaten is waste grain.

THE CROW.

(Corvus americanus.)

There are few birds so well known as the common crow, and unlike most other species he does not seem to decrease in numbers as the country becomes more densely populated. The crow is commonly regarded as a blackleg and a thief. Without the dash and brilliancy of the jay, or the bold savagery of the hawk, he is accused of doing more mischief than either. That he does pull up sprouting corn, destroy chickens, and rob the nests of small birds has been repeatedly proved. Nor are these all of his sins. He is known to eat frogs, toads, salamanders, and some small snakes, all harmless creatures that do some good by eating insects. With so many charges against him, it may be well to show why he should not be utterly condemned.

The examination of a large number of stomachs, while confirming all the foregoing accusations, has thrown upon the subject a light somewhat different from that derived solely from field observation. It shows that the bird's-nesting habit, as in the case of the jay, is not so universal as has been supposed; and that, so far from being a habitual nest robber, the crow only occasionally indulges in that reprehensible practice. The same is true in regard to destroying chickens, for he is able to carry off none but very young ones, and his opportunities for capturing them are somewhat limited. Neither are many toads and frogs eaten, and as frogs are of no great practical value, their destruction is not a serious matter; but toads are very useful, and their consumption, so far as it goes, must be counted against the crow. Turtles, crayfishes, and snails, of which he eats quite a large number, may be considered neutral, while mice may be counted to his credit.

In his insect food, however, the crow makes amends for sins in the rest of his dietary, although even here the first item is against him. Predaceous beetles are eaten in some numbers throughout the season, but the number is not great. May beetles, "dor-bugs," or June bugs, and others of the same family, constitute the principal food during spring and early summer, and are fed to the young in immense quantities. Other beetles, nearly all of a noxious character, are eaten to a considerable extent. Grasshoppers are first taken in May, but not in large numbers until August, when, as might be expected, they form the leading article of diet, showing that the crow is no exception to the general rule that most birds subsist, to a large extent, upon grasshoppers in the month of August. Many bugs, some caterpillars, mostly cutworms, and some spiders are also eaten—all of them either harmful or neutral in their economic relations. Of the insect diet Mr. E. A. Schwarz says: "The facts, on the whole, speak overwhelmingly in favor of the crow."

Probably the most important item in the vegetable food is corn, and by pulling up the newly sprouted seeds the bird renders himself extremely obnoxious. Observation and experiments with tame crows show that hard, dry corn is never eaten if anything else is to be had, and if fed to nestlings it is soon disgorged. The reason crows resort to newly planted fields is that the kernels of corn are softened by the moisture of the earth, and probably become more palatable in the process of germination, which changes the starch of the grain to sugar. The fact, however, remains that crows eat corn extensively only when it has been softened by germination or partial decay, or before it is ripe and still "in the milk." Experience has shown that they may be prevented from pulling up young corn by tarring the seed, which not only saves the corn but forces them to turn their attention to insects. If they persist in eating green corn it is not so easy to prevent the damage; but no details of extensive injury in this way have yet been presented, and it is probable that no great harm has been done.

Crows eat fruit to some extent, but confine themselves for the most part to wild species, such as dogwood, sour gum, and seeds of the different kinds of sumac. They have also a habit of sampling almost everything which appears eatable, especially when food is scarce. For example, they eat frozen apples found on the trees in winter, or pumpkins, turnips, and potatoes which have been overlooked or neglected; even mushrooms are sometimes taken, probably in default of something better.

In estimating the economic status of the crow, it must be acknowledged that he does some damage, but, on the other hand, he should receive much credit for the insects which he destroys. In the more thickly settled parts of the country the crow probably does more good than harm, at least when ordinary precautions are taken to protect young poultry and newly-planted corn against his depredations. If, however, corn is planted with no provision against possible marauders, if hens and turkeys are allowed to nest and to roam with their broods at a distance from farm buildings, losses must be expected.

THE BOBOLINK, OR RICEBIRD.

(*Dolichonyx oryzivorus.*)

The bobolink (fig. 9) is a common summer resident of the United States, north of about latitude 40°, and from New England westward to the Great Plains, wintering beyond our southern border. In New England there are few birds, if any, around which so much romance has clustered; in the South none on whose head so many maledictions have been heaped. The bobolink, entering the United States from the South at a time when the rice fields are freshly sown, pulls up the young plants and feeds upon the seed. Its stay, however, is not long, and it soon hastens northward, where it is welcomed as a herald of summer. During its sojourn in the Northern States it feeds mainly upon insects and small seeds of useless plants; but while rearing its young, insects constitute its chief food, and almost the exclusive diet of its brood. After the young are able to fly, the whole family gathers into a small flock and begins to live almost entirely upon vegetable food. This consists for the most part of weed seeds, since in the North these birds do not appear to molest grain to any great extent. They eat a few oats, but their stomachs do not reveal a great quantity of this or any other grain. As the season advances they gather into larger flocks and move southward, until by the end of August nearly all have left their breeding grounds. On their way they frequent the reedy marshes about the mouths of rivers and on the inland waters of the coast region, subsisting largely upon wild rice. After leaving the Northern States they are commonly known as reed birds, and having become very fat are treated as game.

They begin to arrive on the rice fields in the latter part of August, and during the next month make havoc in the ripening crop. It is unfortunate that the rice districts lie exactly in the track of their fall migration, since the abundant supply of food thus offered has undoubtedly served to attract them more and more, until most of the bobolinks bred in the North are concentrated with disastrous effect on the southeast coast when the rice ripens in the fall. There was evidently a time when no such supply of food awaited the birds on their journey southward, and it seems probable that the introduction of rice culture in the South, combined with the clearing of the forests in the North, thus affording a larger available breeding area, has favored an increase in the

Fig. 9.—Bobolink.

numbers of this species. The food habits of the bobolink are not necessarily inimical to the interests of agriculture. It simply happens that the rice affords a supply of food more easily obtainable than did the wild plants which formerly occupied the same region. Were the rice fields at a distance from the line of migration, or north of the bobolinks' breeding ground, they would probably never be molested; but lying, as they do, directly in the path of migration, they form a recruiting ground, where the birds can rest and accumulate flesh and strength for the long sea flight which awaits them in their course to South America.

The annual loss to rice growers on account of bobolinks has been estimated at $2,000,000. In the face of such losses it is evident that no mere poetical sentiment should stand in the way of applying any remedy which can be devised. It would be unsafe to assume that the insects which the birds consume during their residence in the North can compensate for such destruction. If these figures are any approximation to the truth, the ordinary farmer will not believe that the bobolink benefits the Northern half of the country nearly as much as it damages the Southern half, and the thoughtful ornithologist will be inclined to agree with him. But even if the bird really does more harm than good, what is the remedy? For years the rice planters have been employing men and boys to shoot the birds and drive them away from the fields, but in spite of the millions slain every year their numbers do not decrease. In fact, a large part of the loss sustained is not in the grain which the birds actually eat, but in the outlay necessary to prevent them from taking it all. At present there seems to be no effective remedy short of complete extermination of the species, and this is evidently impracticable even were it desirable.

THE REDWINGED BLACKBIRD.

(Agelaius phœniceus.)

The redwinged, or swamp, blackbird (fig. 10) is found all over the United States and the region immediately to the north. While common in most of its range, its distribution is more or less local, mainly on account of its partiality for swamps. Its nest is built near standing water, in tall grass, rushes, or bushes. Owing to this peculiarity the bird may be absent from large tracts of country which afford no swamps or marshes suitable for nesting. It usually breeds in large colonies, though single families, consisting of a male with several wives, may sometimes be found in a small slough, where each of the females builds her nest and rears her own little brood, while her liege lord displays his brilliant colors and struts in the sunshine. In the Upper Mississippi Valley it finds the conditions most favorable, for the countless prairie sloughs and the margins of the numerous shallow lakes form nesting sites for thousands of redwings; and there are bred the immense flocks which sometimes do so much damage to the grain fields of the West. After the breeding season is over, the birds collect in flocks to migrate, and remain thus associated throughout the winter.

Many complaints have been made against the redwing, and several States have at times placed a bounty upon its head. It is said to cause great damage to grain in the West, especially in the Upper Mississippi Valley; and the rice growers of the South say that it eats rice. No complaints have been received from the Northeastern portion of the country, where the bird is much less abundant than in the West and South.

An examination of 725 stomachs showed that vegetable matter forms 74 per cent of the food, while the animal matter, mainly insects, forms but 26 per cent. A little more than 10 per cent consists of beetles, mostly harmful species. Weevils, or snout beetles, amount to 4 per cent of the year's food, but in June reach 25 per cent. As weevils are among the most harmful insects known, their destruction should condone for at least some of the sins of which the bird has been accused. Grasshoppers constitute nearly 5 per cent of the food, while the rest of the animal matter is made up of various insects, a few snails, and crustaceans. Several dragon flies were found, but these were probably picked up dead, for they are too active to be taken alive, unless by one of the flycatchers. So far as the insect food as a whole is concerned, the redwing may be considered entirely beneficial.

The interest in the vegetable food of this bird centers around the

Fig. 10 —Redwinged blackbird.

grain. Only three kinds, corn, wheat, and oats, were found in appreciable quantities in the stomachs, and they aggregate but little more than 13 per cent of the whole food, oats forming nearly half of this amount. In view of the many complaints that the redwing eats grain, this record is surprisingly small. The crow blackbird has been found to eat more than three times as much. In the case of the crow, corn forms one-fifth of the food, so that the redwinged blackbird, whose diet is made up of only a trifle more than one-eighth of grain, is really one of the least destructive species; but the most important item of this bird's food is weed seed, which forms practically the whole food in winter and about 57 per cent of the whole year's fare. The principal weed seeds eaten are those of ragweed, barn grass, smartweed, and about a dozen others. That these seeds are preferred is shown by the fact that the birds begin to eat them in August, when grain is still readily accessible, and continue feeding on them even after insects become plentiful in April. The redwing eats very little fruit and does practically no harm in the garden or orchard.

While it is impossible to dispute the mass of testimony which has accumulated concerning its grain-eating propensity, the stomach examinations show that the habit must be local rather than general. As the area of cultivation increases and the breeding grounds are curtailed, the species is likely to become reduced in numbers and consequently less harmful. Nearly seven-eighths of the redwing's food is made up of weed seed or of insects injurious to agriculture, indicating unmistakably that the bird should be protected, except, perhaps, in a few places where it is too abundant.

THE MEADOW LARK, OR OLD FIELD LARK.

(Sturnella magna.)

The meadow lark (fig. 11) is a common and well-known bird occurring from the Atlantic Coast to the Great Plains, where it gives way to a closely related subspecies, which extends thence westward to the Pacific. It winters from our southern border as far north as the District of Columbia, southern Illinois, and occasionally Iowa. Although it is a bird of the plains, finding its most congenial haunts in the prairies of the West, it does not disdain the meadows and mowing lands of New England. It nests on the ground and is so terrestrial in its habits that it seldom perches on trees, preferring a fence rail or a telegraph pole. When undisturbed, it may be seen walking about with a peculiar dainty step, stopping every few moments to look about and give its tail a nervous flirt or to sound a note or two of its clear whistle.

The meadow lark is almost wholly beneficial, although a few complaints have been made that it pulls sprouting grain, and one farmer claims that it eats clover seed. As a rule, however, it is looked upon with favor and is not disturbed.

In the 238 stomachs examined, animal food (practically all insects) constituted 73 per cent of the contents and vegetable matter 27 per cent. As would naturally be supposed, the insects were ground species, such as beetles, bugs, grasshoppers, and caterpillars, with a few flies, wasps, and spiders. A number of the stomachs were taken from birds that had been killed when the ground was covered with snow, but still they contained a large percentage of insects, showing the bird's skill in finding proper food under adverse circumstances.

Of the various insects eaten, crickets and grasshoppers are the most important, constituting 29 per cent of the entire year's food and 69 per cent of the food in August. It is scarcely necessary to enlarge upon this point, but it can readily be seen what an effect a number of these birds must have on a field of grass in the height of the grasshopper season. Of the 238 stomachs collected at all seasons of the year, 178, or more than two-thirds, contained remains of grasshoppers, and one was filled with fragments of 37 of these insects. This seems to show conclusively that grasshoppers are preferred and are eaten whenever they can be procured. The great number taken in August is especially noticeable. This is essentially the grasshopper month, i. e., the month when grasshoppers reach their maximum abundance; and the stomach examination has shown that a large number of birds resort to this diet in August, no matter what may be the food during the rest of the year.

Next to grasshoppers, beetles make up the most important item of the meadow lark's food, amounting to nearly 21 per cent, of which about one-third are predaceous ground beetles. The others are all harmful

FIG. 11.—Meadow lark.

species, and when it is considered that the bird feeds exclusively on the ground, it seems remarkable that so few useful ground beetles are eaten. Many of them have a disgusting odor, and possibly this may occasionally save them from destruction by birds, especially when other food is abundant. Caterpillars, too, form a very constant element, and in May constitute over 28 per cent of the whole food. May is the month when the dreaded cutworm begins its deadly career, and then the bird does some of its best work. Most of these caterpillars are ground feeders, and are overlooked by birds which habitually frequent trees; but the meadow lark finds them and devours them by thousands. The remainder of the insect food is made up of a few ants, wasps, and spiders, with a few bugs, including some chinch bugs.

The vegetable food consists of grain, weed, and other hard seeds. Grain in general amounts to 14, and weed and other seeds to 12 per cent.

The grain, principally corn, is mostly eaten in winter and early spring, and must be therefore simply waste kernels; only a trifle is consumed in summer and autumn, when it is most plentiful. No trace of sprouting grain was discovered. Clover seed was found in only six stomachs, and but little in each. Seeds of weeds, principally ragweed, barn grass, and smartweed, are eaten from November to April, inclusive, but during the rest of the year are replaced by insects.

Briefly stated, more than half of the meadow lark's food consists of harmful insects; its vegetable food is composed either of noxious weeds or waste grain, and the remainder is made up of useful beetles or neutral insects and spiders. A strong point in the bird's favor is that, although naturally an insect eater, it is able to subsist on vegetable food, and consequently is not forced to migrate in cold weather any farther than is necessary to find ground free from snow. This explains why it remains for the most part in the United States during winter, and moves northward as soon as the snow disappears from its usual haunts.

There is one danger to which the meadow lark is exposed. As its flesh is highly esteemed the bird is often shot for the table, but it is entitled to all possible protection, and to slaughter it for game is the least profitable way to utilize a valuable species.

THE BALTIMORE ORIOLE.

(*Icterus galbula.*)

Brilliancy of plumage, sweetness of song, and food habits to which no exception can be taken are some of the striking characteristics of the Baltimore oriole (fig. 12). In summer this species is found throughout the northern half of the United States east of the Great Plains, and is welcomed and loved in every country home in that broad land. In the Northern States it arrives rather late, and is usually first seen, or heard, foraging amidst the early bloom of the apple trees, where it searches for caterpillars or feeds daintily on the surplus blossoms. Its nest commands hardly less admiration than the beauty of its plumage or the excellence of its song. Hanging from the tip of the outermost bough of a stately elm, it is almost inaccessible, and so strongly fastened as to bid defiance to the elements.

By watching an oriole which has a nest one may see it searching among the smaller branches of some neighboring tree, carefully examining each leaf for caterpillars, and occasionally trilling a few notes to its mate. Observation both in the field and laboratory shows that caterpillars constitute the largest item of its fare. In 113 stomachs they formed 34 per cent of the food, and are eaten in varying quantities during all the months in which the bird remains in this country, although the fewest are eaten in July, when a little fruit is also taken. The other insects consist of beetles, bugs, ants, wasps, grasshoppers, and some spiders. The beetles are principally click beetles, the larvæ of which are among the most destructive insects known; and the bugs include plant and bark lice, both very harmful, but so small and obscure as to be passed over unnoticed by most birds. Ants are eaten mostly in spring, grasshoppers in July and August, and wasps and spiders with considerable regularity throughout the season.

Vegetable matter amounts to only a little more than 16 per cent of the food during the bird's stay in the United States, so that the possibility of the oriole doing much damage to crops is very limited. The bird has been accused of eating peas to a considerable extent, but remains of peas were found in only two stomachs. One writer says that it damages grapes, but none were found. In fact, a few blackberries and

FIG. 12.—Baltimore oriole.

cherries comprised the only cultivated fruit detected in the stomachs, the remainder of the vegetable food being wild fruit and a few miscellaneous seeds.

THE CROW BLACKBIRD, OR GRACKLE.

(*Quiscalus quiscula.*)

The crow blackbird (fig. 13) or one of its subspecies is a familiar object in all of the States east of the Rocky Mountains. It is a resident throughout the year as far north as southern Illinois, and in summer extends its range into British America. In the Mississippi Valley it is one of the most abundant birds, preferring to nest in the artificial groves and windbreaks near farms instead of the natural "timber" which it formerly used. It breeds also in parks and near buildings, often in considerable colonies. Farther east, in New England, it is only locally abundant, though frequently seen in migration. After July it

becomes very rare, or entirely disappears, owing to the fact that it col-
lects in large flocks and retires to some quiet place, where food is
abundant and where it can remain undisturbed during the molting
season, but in the latter days of August and throughout September it
usually reappears in immense numbers before moving southward.

It is evident that a bird so large and so abundant may exercise an
important influence upon the agricultural welfare of the country it
inhabits. The crow blackbird has been accused of many sins, such as
stealing grain and fruit and robbing the nests of other birds; but the
farmers do not undertake any war of extermination against it, and,
for the most part, allow it to nest about the premises undisturbed. An
examination of 2,258 stomachs showed that nearly one-third of its food
consists of insects, of which the greater part are injurious. The bird

FIG. 13.—Crow blackbird.

also eats a few snails, crayfishes, salamanders, small fish, and occasionally a mouse. The stomach contents do not indicate that it robs other
birds' nests to any great extent, as remains of birds and birds' eggs
amount to less than one-half of 1 per cent.

It is, however, on account of its vegetable food that the grackle is
most likely to be accused of doing damage. Grain is eaten during the
whole year, and during only a short time in summer is other food
attractive enough to induce the bird to alter its diet. The grain taken
in the winter and spring months probably consists of waste kernels
gathered from the stubble. The stomachs do not indicate that the bird
pulls sprouting grain; but the wheat eaten in July and August, and
the corn eaten in the fall, are probably taken from fields of standing

result.

THE SPARROWS.[1]

Sparrows are not obtrusive birds, either in plumage, song, or action. There are some forty species, with nearly as many subspecies, in North America, but their differences, both in plumage and habits, are in most cases too obscure to be readily recognized, and not more than half a dozen forms are generally known in any one locality. All the species are more or less migratory, but so widely are they distributed that there is probably no part of the country where some can not be found throughout the year.

While sparrows are noted seed eaters, they do not by any means confine themselves to a vegetable diet. During the summer, and especially in the breeding season, they eat many insects, and probably feed their young largely upon the same food. An examination of the stomachs of three species—the song sparrow (*Melospiza*), chipping sparrow (*Spizella socialis*), and field sparrow (*Spizella pusilla*) (fig. 14)— shows that about one-third of the food consists of insects, comprising many injurious beetles, such as snout-beetles or weevils, and leaf-beetles. Many grasshoppers are eaten, and in the case of the chipping sparrow these insects form one-eighth of the food. Grasshoppers would seem to be rather large morsels, but the bird probably confines itself to the smaller species; indeed, this is indicated by the fact that the greatest amount (over 36 per cent) is eaten in June, when the larger species are still young and the small species most numerous. Besides the insects already mentioned, many wasps and bugs are taken. Predaceous and parasitic Hymenoptera and predaceous beetles, all useful insects, are

---

[1]The sparrows here mentioned are all native species. For a full account of the English sparrow, including its introduction, habits, and depredations, see Bull. No. 1 of the Division of Ornithology, published in 1889.

eaten only to a slight extent, so that as a whole the sparrows' insect diet may be considered beneficial.

Their vegetable food is limited almost exclusively to hard seeds This might seem to indicate that the birds feed to some extent upon grain, but the stomachs examined show only one kind—oats—and but little of that. The great bulk of the food is made up of grass and weed seed, which form almost the entire diet during winter, and the amount consumed is immense.

Anyone acquainted with the agricultural region of the Upper Mississippi Valley can not have failed to notice the enormous growth of weeds in every waste spot where the original sward has been disturbed.

FIG. 14 —Field sparrow.

By the roadside, on the borders of cultivated fields, or in abandoned fields, wherever they can obtain a foothold, masses of rank weeds spring up, and often form impenetrable thickets which afford food and shelter for immense numbers of birds and enable them to withstand great cold and the most terrible blizzards. A person visiting one of these weed patches on a sunny morning in January, when the thermometer is 20° or more below zero, will be struck with the life and animation of the busy little inhabitants. Instead of sitting forlorn and half frozen, they may be seen flitting from branch to branch, twittering and fluttering, and showing every evidence of enjoyment and perfect comfort. If one of them be killed and examined, it will be found in excellent condition—in fact, a veritable ball of fat.

The snowbird (*Junco hyemalis*) and tree sparrow (*Spizella monticola*) are perhaps the most numerous of all the sparrows. The latter fairly swarms all over the Northern States in winter, arriving from the north early in October and leaving in April. Examination of many stomachs shows that in winter the tree sparrow feeds entirely upon seeds of weeds; and probably each bird consumes about one-fourth of an ounce a day. In an article contributed to the New York Tribune in 1881 the writer estimated the amount of weed seed annually destroyed by these birds in the State of Iowa. Upon the basis of one-fourth of an ounce of seed eaten daily by each bird, and supposing that the birds averaged ten to each square mile, and that they remain in their winter range two hundred days, we shall have a total of 1,750,000 pounds, or 875 tons, of weed seed consumed by this one species in a single season. Large as these figures may seem, they certainly fall far short of the reality. The estimate of ten birds to a square mile is much within the truth, for the tree sparrow is certainly more abundant than this in winter in Massachusetts, where the food supply is less than in the Western States, and I have known places in Iowa where several thousand could be seen within the space of a few acres. This estimate, moreover, is for a single species, while, as a matter of fact, there are at least half a dozen birds (not all sparrows) that habitually feed on these seeds during winter.

Farther south the tree sparrow is replaced in winter by the white-throated sparrow, the white-crowned sparrow, the fox sparrow, the song sparrow, the field sparrow, and several others; so that all over the country there are a vast number of these seed eaters at work during the colder months reducing next year's crop of worse than useless plants.

In treating of the value of birds, it has been customary to consider them mainly as insect destroyers; but the foregoing illustration seems to show that seed eaters have a useful function, which has never been fully appreciated.

THE ROSE-BREASTED GROSBEAK.

(*Zamelodia ludoviciana.*)

The beautiful rose-breasted grosbeak (fig. 15) breeds in the northern half of the United States east of the Missouri River, but spends its winters beyond our boundaries. Unfortunately it is not abundant in New England, and nowhere as plentiful as it should be. It frequents groves and orchards rather than gardens or dooryards, but probably the beauty of the male is the greatest obstacle to its increase; the fully adult bird is pure black and white, with a broad patch of brilliant rose color upon the breast and under each wing. On account of this attractive plumage the birds are highly prized for ladies' hats; and consequently have been shot in season and out, till the wonder is not that there are so few, but that any remain at all.

When the Colorado potato beetle first swept over the land, and naturalists and farmers were anxious to discover whether or not there were any enemies which would prey upon the pest, the grosbeak was almost the only bird seen to eat the beetles. Further observation confirmed the fact, and there can be no reasonable doubt that where the bird is abundant it has contributed very much to the abatement of the pest which has been noted during the last decade. But this is not the only good which the bird does, for many other noxious insects besides the potato beetle are also eaten.

The vegetable food of the grosbeak consists of buds and blossoms of forest trees, and seeds, but the only damage of which it has been

Fig. 15.—Rose-breasted grosbeak.

accused is the stealing of green peas. The writer has observed it eating peas and has examined the stomachs of several that had been killed in the very act. The stomachs contained a few peas and enough potato beetles, old and young, as well as other harmful insects, to pay for all the peas the birds would be likely to eat in a whole season. The garden where this took place adjoined a small potato field which earlier in the season had been so badly infested with the beetles that the vines were completely riddled. The grosbeaks visited the field every day, and finally brought their fledged young. The young birds stood in a row on the topmost rail of the fence and were fed with the beetles which their parents gathered. When a careful inspection was made a few days later, not a beetle, old or young, could be found; the birds had swept them from the field and saved the potatoes.

There are seven common species of swallows within the limits of the United States, four of which have, to some extent, abandoned their

Fig 16 —Barn swallow

primitive nesting habits and attached themselves to the abodes of man. As a group, swallows are gregarious and social in an eminent degree. Some species build nests in large colonies, occasionally numbering thousands; in the case of others only two or three pairs are found together; while still others nest habitually in single pairs.

Their habits are too familiar to require any extended description. Their industry and tirelessness are wonderful, and during the day it is rare to see swallows at rest except just before their departure for the South, when they assemble upon telegraph wires or upon the roofs of buildings, apparently making plans for the journey.

A noticeable characteristic of several of the species is their attachment to man. In the eastern part of the country the barn swallow (*Chelidon erythrogastra*) (fig. 16) now builds exclusively under roofs, having entirely abandoned the rock caves and cliffs in which it formerly nested. More recently the cliff swallow (*Petrochelidon lunifrons*) has found a better nesting site under the eaves of buildings than was afforded by the overhanging cliffs of earth or stone which it once used, and to which it still resorts occasionally in the East, and habitually in the unsettled West. The martin (*Progne subis*) and white-bellied swallow (*Tachycineta bicolor*) nest either in houses supplied for the purpose, in abandoned nests of woodpeckers, or in natural crannies in rocks. The other species have not yet abandoned their primitive habitats, but possibly may do so as the country becomes more thickly settled.

Field observation will convince any ordinarily attentive person that the food of swallows must consist of the smaller insects captured in mid-air, or perhaps in some cases picked from the tops of tall grass or weeds. This observation is borne out by an examination of stomachs, which shows that the food consists of many small species of beetles which are much on the wing; many species of Diptera (mosquitoes and their allies), with large quantities of flying ants and a few insects of similar kinds. Most of them are either injurious or annoying, and the numbers destroyed by swallows are not only beyond calculation, but almost beyond imagination.

The white-bellied swallow eats a considerable number of berries of the bayberry, or wax myrtle. During migrations and in winter it has a habit of roosting in these shrubs, and it probably obtains the fruit at that time.

It is a mistake to tear down the nests of a colony of cliff swallows from the caves of a barn, for so far from disfiguring a building the nests make a picturesque addition, and their presence should be encouraged by every device. It is said that cliff and barn swallows can be induced to build their nests in a particular locality, otherwise suitable, by providing a quantity of mud to be used as mortar. Barn swallows may also be encouraged by cutting a small hole in the gable of the barn, while martins and white-bellied swallows will be grateful for boxes like those for the bluebird, but placed in some higher situation.

THE CEDAR BIRD.

(*Ampelis cedrorum.*)

The cedar waxwing, or cherry bird (fig. 17), inhabits the whole of the United States, but is much less common in the West. Although the great bulk of the species retires southward in winter, the bird is occasionally found in every State during the colder months, especially if wild berries are abundant. Its proverbial fondness for cherries has given rise to its popular name, and much complaint has been made on account of the fruit eaten. Observation has shown, however, that its depredations are confined to trees on which the fruit ripens earliest, while later varieties are comparatively untouched. This is probably owing to the fact that when wild fruits ripen they are preferred to cherries, and really constitute the bulk of the cedar bird's diet.

In 152 stomachs examined animal matter formed only 13 and vegeta- ble 87 per cent, showing that the bird is not wholly a fruit eater. With the exception of a few snails, all the animal food consisted of insects, mainly beetles—and all but one more or less noxious, the famous elm leaf-beetle being among the number. Bark or scale lice were found in several stomachs, while the remainder of the animal food was made up of grasshoppers, bugs, and the like. Three nestlings were found to have been fed almost entirely on insects.

Of the 87 per cent of vegetable food, 74 consisted entirely of wild fruit or seeds and 13 of cultivated fruit, but a large part of the latter

FIG. 17 —Cedar bird.

was made up of blackberries and raspberries, and it is very doubtful whether they represented cultivated varieties. Cherry stealing is the chief complaint against this bird, but of the 152 stomachs only 9, all taken in June and July, contained any remains of cultivated cherries, and these aggregate but 5 per cent of the year's food. As 41 stomachs were collected in those months, it is evident that the birds do not live to any great extent on cultivated cherries.

Although the cherry bird is not a great insect destroyer, it does some good work in this way, since it probably rears its young mostly upon insect food. On the other hand, it does not devour nearly as much cultivated fruit as has been asserted, and most, if not all, of the damage can be prevented. The bird should therefore be considered a useful species, and as such should be accorded all possible protection.

THE CATBIRD.

(Galeoscoptes carolinensis.)

The catbird (fig. 18), like the thrasher, is a lover of swamps, and delights to make its home in a tangle of wild grapevines, greenbriers, and shrubs, where it is safe from attack and can find its favorite food in abundance. It is found throughout the United States west to the Rocky Mountains; occurs also in Washington, Idaho, and Utah, and extends northward into British America. It winters in the Southern States, Cuba, Mexico, and Central America.

The catbird always attracts attention, and the intruder upon its haunts soon understands that he is not welcome. There is no mistaking the meaning of the sneering voice with which he is saluted, and there is little doubt that this gave rise to the popular prejudice against the

Fɪɢ. 18.—Catbird.

bird; but the feeling has been increased by the fact that the species is sometimes a serious annoyance to fruit growers. All such reports, however, seem to come from the prairie country of the West. In New England, according to the writer's experience the catbird is seldom seen about gardens or orchards; the reason may possibly be found in the fact that on the prairies fruit-bearing shrubs which afford so large a part of this bird's food are conspicuously absent. With the settlement of this region comes an extensive planting of orchards, vineyards, and small fruit gardens, which furnish shelter and nesting sites for the catbird, as well as for other species, with a consequent large increase in their numbers, but without providing the native fruits upon which they have been accustomed to feed. Under these circumstances, what is more natural than for the birds to turn to cultivated fruits for their supplies? The remedy is obvious; cultivated fruits can be protected by the simple expedient of planting wild species or others which are preferred by the birds. Some experiments with catbirds in captivity showed that the Russian mulberry was preferred to any cultivated fruit that could be offered.

The stomachs of 213 catbirds were examined and found to contain 44 per cent of animal (insect) and 56 per cent of vegetable food.[1] Ants, beetles, caterpillars, and grasshoppers constitute three-fourths of the animal food, the remainder being made up of bugs, miscellaneous insects and spiders. One-third of the vegetable food consists of cultivated fruits, or those which may be cultivated, such as strawberries, raspberries, and blackberries; but while we debit the bird with the whole of this, it is probable—and in the eastern and well-wooded part of the country almost certain—that a large part was obtained from wild vines. The rest of the vegetable matter is mostly wild fruit, such as cherries, dogwood, sour gum, elder berries, greenbrier, spice berries, black alder, sumac, and poison ivy.

Although the catbird sometimes does considerable harm by destroying small fruit, the bird can not be considered injurious. On the contrary, in most parts of the country it does far more good than harm, and the evil it does can be reduced appreciably by the methods already pointed out.

THE BROWN THRASHER.

(*Harporhynchus rufus.*)

The brown thrasher (fig. 19) breeds throughout the United States east of the Great Plains, and winters in the south Atlantic and Gulf States. It occasionally visits the garden or orchard, but nests in swamps or in groves standing upon low ground. While it generally prefers a thickly grown retreat, it sometimes builds in a pile of brush at a distance from trees. On account of its more retiring habits it is not so conspicuous as the robin, although it may be equally abundant. Few birds can excel the thrasher in sweetness of song, but it is so shy that its notes are not heard often enough to be appreciated. Its favorite time for singing is the early morning, when, perched on the top of some tall bush or low tree, it gives an exhibition of vocal powers which would do credit to a mockingbird. Indeed, in the South, where the latter bird is abundant, the thrasher is known as the sandy mocker.

The food of the brown thrasher consists of both fruit and insects. An examination of 121 stomachs showed 36 per cent of vegetable and 64 of animal food, practically all insects, and mostly taken in spring before fruit is ripe. Half the insects were beetles, and the remainder chiefly grasshoppers, caterpillars, bugs, and spiders. A few predaceous

[1] The investigation of the food of the catbird, brown thrasher, and house wren was made by Mr. Sylvester D. Judd and published in the Yearbook of the Department of Agriculture for 1895, pp. 405–408.

beetles were eaten, but, on the whole, its work as an insect destroyer may be considered beneficial.

Eight per cent of the food is made up of fruits like raspberries and currants which are or may be cultivated, but the raspberries at least are as likely to belong to wild as to cultivated varieties. Grain, made up mostly of scattered kernels of oats and corn, is merely a trifle, amounting to only 3 per cent, and though some of the corn may be taken from newly planted fields it is amply paid for by the May beetles which are eaten at the same time. The rest of the food consists of wild fruit or seeds. Taken all in all, the brown thrasher is a useful bird, and probably does just as good work in its secluded retreats as

Fig. 19.—Brown thrasher.

it would about the garden, for the swamps and groves are no doubt the breeding grounds of many insects that migrate thence to attack the farmers' crops.

THE HOUSE WREN.

(*Troglodytes aëdon.*)

The diminutive house wren (fig. 20) frequents barns and gardens, and particularly old orchards in which the trees are partially decayed. He makes his nest in a hollow branch where perhaps a woodpecker had a domicile the year before, but he is a pugnacious character, and if he happens to fancy one of the boxes that have been put up for the bluebirds he does not hesitate to take it. He is usually received with favor, and is not slow to avail himself of boxes, gourds, tin cans, or empty jars placed for his accommodation.

As regards food habits, the house wren is entirely beneficial. Practically, he can be said to live upon animal food alone, for an examination

. of 52 stomachs showed that 98 per cent of the stomach contents was made up of insects or their allies, and only 2 per cent was vegetable, including bits of grass and similar matter, evidently taken by accident with the insects. Half of this food consisted of grasshoppers and beetles; the remainder of caterpillars, bugs, and spiders. As the house wren is a prolific breeder, frequently rearing from twelve to sixteen young in a season, a family of these birds must cause considerable reduction in the number of insects in a garden. Wrens are industrious foragers, searching every tree, shrub, or vine for caterpillars, examining every post and rail of the fence, and every cranny in the wall for insects or spiders. They do not, as a rule, fly far afield, but work

Fig. 20.—House wren.

industriously in the immediate vicinity of their nests. In this way they become valuable aids in the garden or orchard, and by providing suitable nesting boxes they may be induced to take up residence where their services will do most good. Their eccentricities in the selection of a home are well known. Almost anything, from an old cigar box to a tomato can, an old teapot, a worn-out boot, or a horse's skull, is acceptable, provided it be placed well up from the ground and out of reach of cats and other prowlers.

It does not seem possible to have too many wrens, and every effort should be made to protect them and to encourage their nesting about the house.

THE ROBIN.

(Merula migratoria.)

The robin (fig. 21) is found throughout the United States east of the Great Plains, and is represented farther west by a slightly different subspecies. It extends far north through Canada, and is found even in Alaska. Although the great bulk of the species leaves the Northern States in winter, a few individuals remain in sheltered swamps, where wild berries furnish an abundant supply of food.

The robin builds its nest in orchards and gardens, and occasionally takes advantage of a nook about the house, or under the shelter of the roof of a shed or outbuilding. Its food habits have sometimes caused apprehension to the fruit grower, for it is fond of cherries and other small fruits, particularly the earlier varieties. For this reason many complaints have been lodged against it, and some persons have gone

Fig. 21.—Robin.

so far as to condemn the bird. The robin is, however, too valuable to be exterminated, and choice fruit can be readily protected from its depredations.

An examination of 330 stomachs shows that over 42 per cent of its food is animal matter, principally insects, while the remainder is made up largely of small fruits or berries. Over 19 per cent consists of beetles, about one-third of which are useful ground beetles, taken mostly in spring and fall, when other insects are scarce. Grasshoppers make up about one-tenth of the whole food, but in August comprise over 30 per cent. Caterpillars form about 6 per cent, while the rest of the animal food, about 7 per cent, is made up of various insects, with a few spiders, snails, and angle-worms. All the grasshoppers, caterpillars, and bugs, with a large portion of the beetles, are injurious, and it is safe to say that noxious insects comprise more than one-third of the robin's food.

Vegetable food forms nearly 58 per cent of the stomach contents, over 47 being wild fruits, and only a little more than 4 per cent being possibly cultivated varieties. Cultivated fruit amounting to about 25 per cent was found in the stomachs in June and July, but only a trifle in August. Wild fruit, on the contrary, is eaten in every month, and constitutes a staple food during half the year. No less than forty-one species were identified in the stomachs; of these, the most important were four species of dogwood, three of wild cherries, three of wild grapes, four of greenbrier, two of holly, two of elder; and cranberries, huckleberries, blueberries, barberries, service berries, hackberries, and persimmons, with four species of sumac, and various other seeds not strictly fruit.

The depredations of the robin seem to be confined to the smaller and earlier fruits, and few, if any, complaints have been made against it on the score of eating apples, peaches, pears, grapes, or even late cherries. By the time these are ripe the forests and hedges are teeming with wild fruits, which the bird evidently finds more to its taste. The cherry, unfortunately, ripens so early that it is almost the only fruit accessible at a time when the bird's appetite has been sharpened by a long-continued diet of insects, earthworms, and dried berries, and it is no wonder that at first the rich juicy morsels are greedily eaten. In view of the fact that the robin takes ten times as much wild as cultivated fruit, it seems unwise to destroy the birds to save so little. Nor is this necessary, for by a little care both may be preserved. Where much fruit is grown, it is no great loss to give up one tree to the birds; and in some cases the crop can be protected by scarecrows. Where wild fruit is not abundant, a few fruit-bearing shrubs and vines judiciously planted will serve for ornament and provide food for the birds. The Russian mulberry is a vigorous grower and a profuse bearer, ripening at the same time as the cherry, and, so far as observation has gone, most birds seem to prefer its fruit to any other. It is believed that a number of these trees planted around the garden or orchard would fully protect the more valuable fruits.

Many persons have written about the delicate discrimination of birds for choice fruit, asserting that only the finest and costliest varieties are selected. This is contrary to all careful scientific observation. Birds, unlike human beings, seem to prefer fruit like the mulberry, that is sweetly insipid, or that has some astringent or bitter quality like the chokecherry or holly. The so-called black alder (*Ilex verticillata*), which is a species of holly, has bright scarlet berries, as bitter as quinine, that ripen late in October, and remain on the bushes through November, and though frost grapes, the fruit of the Virginia creeper, and several species of dogwood are abundant at the same time, the birds eat the berries of the holly to a considerable extent, as shown by the seeds found in the stomachs. It is moreover a remarkable fact that the wild fruits upon which the birds feed largely are those which man neither gathers for his own use nor adopts for cultivation.

THE BLUEBIRD.

(Sialia sialis.)

The common and familiar bluebird (fig. 22) is an inhabitant of all the States east of the Rocky Mountains from the Gulf of Mexico north ward into Canada. It winters as far north as southern Illinois, in the Mississippi Valley, and Pennsylvania in the east; in spring it is one of the first migrants to arrive in the Northern States, and is always welcomed as an indication of the final breaking up of winter. It frequents orchards and gardens, where it builds its nest in hollow trees, or takes advantage of a nesting box provided by the enterprising farmer's boy.

So far as known, this bird has not been accused of stealing fruit or of preying upon any crops. An examination of 205 stomachs showed

Fig. 22.—Bluebird.

that 76 per cent of the food consists of insects and their allies, while the other 24 per cent is made up of various vegetable substances, found mostly in stomachs taken in winter. Beetles constitute 28 per cent of the whole food, grasshoppers 22, caterpillars 11, and various insects, including quite a number of spiders, comprise the remainder of the insect diet. All these are more or less harmful, except a few predaceous beetles, which amount to 8 per cent, but in view of the large consumption of grasshoppers and caterpillars, we can at least condone this offense, if such it may be called. The destruction of grasshoppers is very noticeable in the months of August and September, when these insects form more than 60 per cent of the diet.

It is evident that in the selection of its food the bluebird is governed more by abundance than by choice. Predaceous beetles are eaten in spring, as they are among the first insects to appear; but in early summer caterpillars form an important part of the diet, and are replaced a little later by grasshoppers. Beetles are eaten at all times, except when grasshoppers are more easily obtained.

So far as its vegetable food is concerned, the bluebird is positively harmless. The only trace of any useful product in the stomachs consisted of a few blackberry seeds, and even these more probably belonged to wild than cultivated varieties. Following is a list of the various seeds which were found: Blackberry, chokeberry, juniperberry, poke, berry, partridgeberry, greenbriar, Virginia creeper, bittersweet, holly-strawberry bush, false spikenard, wild sarsaparilla, sumac (several species), rose haws, sorrel, ragweed, grass, and asparagus. This list shows how little the bluebird depends upon the farm or garden to supply its needs, and indicates that by encouraging the growth of some of these plants, many of which are highly ornamental, the bird can be induced to make its home on the premises.

Bluebirds are so well known that it seems unnecessary to urge anything more in their favor; but in view of the fact that large numbers were destroyed during the severe storm of 1895, more than ordinary vigilance should be exercised in protecting them until they have regained their normal abundance.

○

Lightning Source UK Ltd.
Milton Keynes UK
UKHW021039110119
335297UK00012B/1734/P